Ângela Neta Dias dos Santos
Mariluce Messias

Birds as indicators of environmental quality

AF666066

Ângela Neta Dias dos Santos
Mariluce Messias

Birds as indicators of environmental quality

Studies in forest fragments in Rondônia

ScienciaScripts

Imprint

Cover image: www.ingimage.com

This book is a translation from the original published under ISBN 978-3-330-75853-7.

Publisher:
Sciencia Scripts
is a trademark of
Dodo Books Indian Ocean Ltd. and OmniScriptum S.R.L publishing group

120 High Road, East Finchley, London, N2 9ED, United Kingdom
Str. Armeneasca 28/1, office 1, Chisinau MD-2012, Republic of Moldova, Europe
Printed at: see last page
ISBN: 978-620-8-30226-9

SUMMARY

DEDICATORY

I dedicate this monograph to my family and my boyfriend Diego Meneguelli"

ACKNOWLEDGMENTS

To my mother, Maria da Glòria Dias de Jesus, who, even though she was far away, never stopped believing in my potential and always gave me the strength I needed along the way.

To my siblings, who gave me all their love and attention, as well as many moments of relaxation.

To my boyfriend, Diego Meneghelli, for his great help: tips and corrections to this work, being a second "advisor" to me. He was also the person who always believed in making my dreams come true. But, above all, I thank you for all your love, your unconditional companionship, your understanding and your strength in times of discouragement. Thank you so much! I love you very much!

To my friends, who are considered brothers, Elvis Brambilla (rabid Pinscher), Liria Daiane (Pé de coco), Amanda Teixeira (A preta) and Ravena Mendonça, who at times when the financial situation was difficult, she always made "soup for us" (lol). All of them have been very present in my life during these four years. I'll always be grateful for everything they've done for me.

To my advisor, Dr. Mariluce Rezende Messias, who enabled me to carry out this work.

SUMMARY

The process of deforestation in forest areas can lead to the formation of isolated fragments, which for most birds is a very important factor, since many species are unable to move to other fragments that are too far away, which prevents the flow of individuals between fragments and may, over time, reduce the genetic variability of populations. The main objective of this study was to assess the conservation status of four forest fragments using the trophic guilds of the birds recorded as bioindicators. The fragments were named F1 (approximately 335 ha), F2 (approximately 217 ha), F3 (approximately 538 ha), located on the right bank of the Madeira River, and F4 (approximately 308 ha), located on the left bank of the Madeira River, all in the municipality of Porto Velho/RO. The study was carried out between January and February 2013 and used the observation point method. The sampling effort lasted 14 days, totaling 112 effective hours of observation. 210 points of 20 minutes each were made. The minimum spacing between points was 100 meters. Observations took place in the morning and late afternoon, using binoculars and a digital camera. Considering all the environments studied, 133 species of birds were recorded, distributed in 31 families and 19 orders. The best represented trophic guilds were omnivores and insectivores, both with 24% of the total species recorded in all the fragments analyzed. In general, it can be considered that these fragments have reduced environmental quality, as these guilds, especially the omnivores, represent species that are able to adapt to different diets in different environmental conditions, which makes them increasingly abundant in fragmented environments.

Keywords:

Avifauna, trophic guilds, bioindicators, fragmentation, Rondônia.

INTRODUCTION

The Amazon is the largest tropical rainforest in the world and covers a large part of the northern region of Brazil, occupying almost half (49.29%) of the Brazilian territory, one of the reasons why Brazil is a champion in biodiversity, topping the list of megadiverse countries (VIEIRA, 2006). This biome completely covers five states (Acre, Amapâ, Amazonas, Parâ and Roraima), almost entirely Rondônia (98.8%) and partially the states of Mato Grosso (54%), Maranhao (34%) and Tocantins (9%) (AMBIENTE BRASIL, 2011). It is dominated by a hot and humid climate (with an average temperature of 25 °C) and by forests. It has torrential rainfall well distributed throughout the year and rivers with intense flow (AMBIENTE BRASIL, 2011).

The process of occupying the Amazon has triggered a series of factors that have led to deforestation, including the expansion of cattle ranching and the agricultural frontier, the timber industry, colonization projects, the opening of new roads and other infrastructure, the system of granting tenure rights, population growth, land speculation and regional economic growth, among others (ANGELO & SA, 2007; MARGULIS, 2003).

The causes of accelerated deforestation in the Amazon are multiple, including, recently, the construction of large hydroelectric dams (pers. obs.).

Considering the accumulated deforestation in the six months of the current deforestation calendar (August 2012 to January 2013), Pará leads the ranking with 49% of the total deforested. This is followed by Mato Grosso with 23%, Rondônia with 13% and Amazonas with 12%. These four states were responsible for 97% of the deforestation in the Legal Amazon during this period (IMAZON, 2013).

In Rondônia, this process of deforestation began with the opening of the BR 364 highway, Cuiabâ/Porto Velho. Since 1960, the state's population has grown faster than the national average. This growth was vigorous at times, such as between 1970 and 1980, when its population grew by 16%, more than five times the national average and more than three times the average for the Northern Region (IBGE, 2008). All this population growth was mainly due to the migratory movements of settlers from the South in search of land for agriculture (FEARNSIDE, 2005; 2006).

However, with inadequate soil management, difficulties in accessing the market and low yields, almost all of the planted area was converted into pasture to house a growing cattle herd, which today numbers around 12 million head and occupies almost 6 million hectares planted with grasses (SEDAM, 2009).

Deforestation was on the rise until 2005, and only since then has there been a drop in the rates (Figure 1) (INPE, 2012). However, it should be borne in mind that in many of Rondônia's municipalities the deforested area has already exceeded 80% of the municipal territory (INPE, 2007).

Figure 1. Comparison of the rate of deforestation between Rondônia and the states with the highest rates of deforestation in the Legal Amazon: Mato Grosso and Para. Consolidated rates 1988 to 2012 (km2/year) (INPE, 2012).

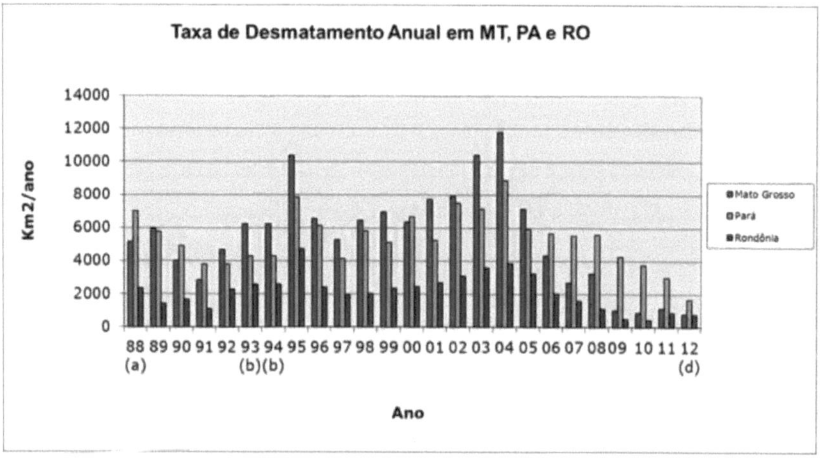

Of all the municipalities in the state of Rondônia, Porto Velho had the highest rate of deforestation up to 2011 (Figure 2), and is the site of the construction of two hydroelectric power stations. The areas close to these constructions have become the focus of occupation by loggers and family farmers, with a significant increase in deforestation in recent years (pers. obs.).

Figura 2. Deforestation situation for the 10 most deforested municipalities in Rondônia from 2000 to 2011 (INPE, 2012). Source:< http://www.dpi.inpe.br/prodesdigital/prodesmunicipal.php> Accessed: February 2013.

Distribution of the 86114.8 km2 of DEFORESTATION up to 2011 in RO

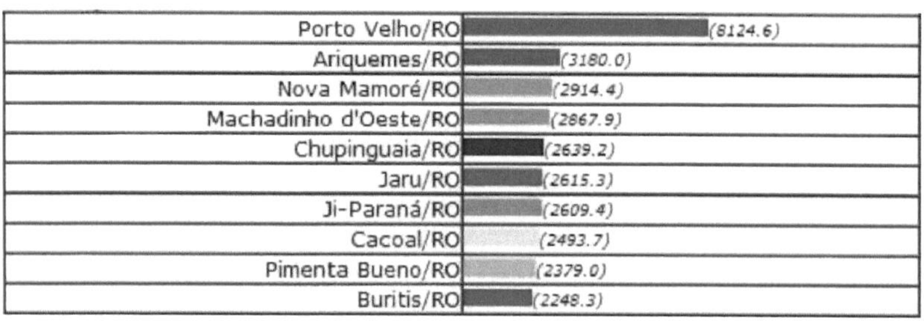

Another important point is that in 2010 construction began on the bridge over the Madeira River that will connect Rondônia to the states of Amazonas and Roraima via the BR-319 highway, which is scheduled for completion in early 2013. This will facilitate traffic to the left bank of the river, where there is already a process of real estate speculation and urban expansion on this bank. Because of this, it is certain that the rate of deforestation in this region will be high in the coming years (pers. obs.).

In addition, the City Council (Concidade), in an extraordinary meeting, approved the change proposed by the municipal executive in the City Master Plan, so that the municipality can build the houses for the families that will be relocated because of the construction of the Madeira River bridge (pers. obs.).

With the council's approval of the change, the executive will submit a supplementary bill to the city council to extend the municipality's urban perimeter, so that the area on the other side of the river becomes part of the urban perimeter. The bill declares the area on the left bank of the Madeira River to be an Urban Expansion Zone.

The project also provides for the creation of a Special Social Interest Zone (ZEIS) and the preparation of a Social Interest Land Regularization Project, taking into account the characteristics of the occupation and the area to be occupied, where specific urban and environmental parameters will be defined, In addition to identifying the plots, the circulation routes and the areas destined for public use (JORNAL GENTE DE OPINIÂO, 2012), all of this will mean that the municipality will lose large areas, which until then had been forested along the left bank of the Madeira River.

A direct and inevitable consequence of deforestation in the municipality of Porto Velho is forest fragmentation, which is a widespread phenomenon associated with the expansion of human development frontiers (VIANA *et al.*, 1997). The rate at which man is altering natural landscapes is thousands of times greater than the natural disturbance dynamics of ecosystems (TABARELLI & GASCON, 2004).

Fragmentation is a form of cover in which large areas of forest are replaced by irregular, usually asymmetrical mosaics of remaining forest and the vegetation cover that the forest has been replaced by (LAURENCE & BIERREGAARD, 1997; SKOLE & TUCKER, 1997). BIERREGAARD, 1997), which is one of the characteristic landscape patterns of tropical forests today (LAURENCE & BIERREGAARD, 1997; SKOLE & TUCKER, 1993; TABARELLI & GASCON, 2004). This fragmentation of the landscape directly affects the ecological (LAURENCE *et al.,* 1997), hydrological and climatic processes (AVISSAR & PIELKE, 1989; GIAMBELLUCA *et al.*, 2003; KAPOS, 1989) of ecosystems.

The process of deforestation in forest areas can lead to the formation of isolated fragments that function as "islands" of forest surrounded by non-forested habitats (PÉRICO *et al.*, 2005). For this reason, the study of habitat fragmentation was initially based on MacArthur & Wilson's (1967) Theory of Island Biogeography. The basic foundation of this theory is that islands of different sizes are home to different numbers of species, due to both the carrying capacity of the environment and the probability of colonization by species, based on the balance between extinction and immigration

(MacARTHUR & WILSON, 1967).

Thus, the theory of island biogeography (MacARTHUR & WILSON, 1967) analyzed the determinants of species richness on islands and emphasized that many of the principles observed in remote archipelagos apply to island habitats on the mainland. Thus, the similarity between islands and forest fragments surrounded by an environment modified by anthropogenic action, stimulated further research applying some of the principles of island biogeography to explain species richness in forest fragments (LAPS, 2006). However, Warburton (1997) and Gimenes & Anjos (2003) pointed out that there are criticisms of applying the theory for this purpose, because among other problems, it does not provide information on the composition of species that fragmented habitats can support.

As the theory of island biogeography was initially proposed for islands, the isolation factor would therefore have a stronger influence on the process than in forest fragments (GIMENES & ANJOS, 2003). However, according to Terborgh *et al.* (1997) the degree of isolation affects the number of species not only on islands. Forest fragments surrounded by a matrix habitat that is very different from the forest vegetation and isolated from other forests usually show isolation effects similar to those seen on islands. In the case of birds, the degree of isolation is a very important factor, as many species are unable to move to other fragments that are too far away, especially those with low flight capacity (pers. obs.).

In addition, forest clearings around the fragment can represent a barrier for many species of birds adapted to living in the interior of forests, which prevents the flow of individuals between fragments and can, over time, reduce the genetic variability of these populations. In fact, abrupt transitions between different types of vegetation represent a barrier to bird dispersal (HAYES, 1995).

The physiognomy of the vegetation is also a very important factor in determining the richness of the avifauna in a fragment because, according to Holmes (1990a), it influences the ease or success with which birds can obtain their resources and the greater the structural variety found in the vegetation of a forest, the greater the possibility of a greater number of bird species finding suitable substrates to optimize their fundamental activities, such as foraging and nesting.

Thus, the way in which birds obtain their resources differs between species and between microhabitats and a bird can have a different level of success depending on the substrate and stratum in which it is operating (HOLMES, 1990a). Thus, different plant species and different vegetation patterns correspond to different microhabitats for birds in a forest (HOLMES, 1990b). Fragmentation can break up this variety of microhabitats that previously existed in the forest, causing some microhabitats to remain in a fragment and others to disappear. Species that need a specific microhabitat may disappear in fragments where it no longer exists and species that periodically require different microhabitats, which are now present in different fragments, may be unable to reach them due to the barriers caused by fragmentation (WILCOVE & ROBINSON, 1990).

Various ecological indicators have been used to analyze changes in the environment and in environmental impact studies to detect changes in the condition of the environment (ANDRADE *et al.*, 2011). Within this context, wild birds are recognized as the best bioindicators of terrestrial ecosystems, especially forest ecosystems (ANJOS, 2001). Bird species occupy many ecological and trophic niches in forests, ranging from the ground to the treetops, which makes it possible to analyze these environments (ALMEIDA, 1998).

A trophic guild is defined as a group of species that exploit the same class of environmental resources in a similar way (ROOT, 1967). In a more restricted sense, a guild would be a group of species that feed on the same food resource in similar proportions (POULIN *et al.*, 1994b; SIMBERLOFF & DAYAN 1991).

Different groups of birds respond in different ways to changes in the environment. And these observations can be made using different guilds as indicators of environmental quality in forest fragments, as the presence of birds from the large frugivore guild is a very relevant aspect and indicates the good quality of the environment (ALEIXO & VIELLIARD, 1995; DARIO, 2008), since birds of this type are very demanding with regard to their food, requiring a high density of fruit in the forest to make up their diet, not least because birds of this size are not able to cross very extensive open fields in search of food, especially when *there are* no *trees* in the adjacent matrix that can serve as "stepping stones". In addition, the species in this guild favor the maintenance of forest fragments, being responsible for a large part of seed dispersal, contributing to their conservation.

Another aspect that is well considered is the presence of carnivores in forest fragments, as these indicate the presence of relatively balanced ecosystems, since carnivore species are usually predators at the top of the food chain, also needing large areas to forage and are harmed by degradation and changes in the abundance of their prey (ALEIXO, 1999).

The guild of "branch and trunk insectivores", made up of the Picidae and Dendrocolaptidae families, species that require cavities for reproduction (in the same way as the Psittacidae), when it has representatives of several species, indicates a good quality forest environment. These birds are specialists and according to Sick (1997), one cannot generalize when it comes to the feeding habits of these groups.

The diets of the Picidae and Dendrocolaptidae are mainly composed of insect larvae found in standing dead trees. The scarcity of these trees is a limiting factor for species of Picidae, Dendrocolaptidae and Psitacidae, whose forest species nest exclusively in the hollows of sticks (SICK, 1997).

On the other hand, within the aspect of the guilds found in an ecosystem, a worrying fact is the record of high rates of omnivores, which are species with high plasticity in terms of the impacts caused by human activities in altered landscapes and a high capacity to adapt to altered environments, i.e. they are undemanding in terms of environmental resources (ALEIXO, 1999).

In this way, the specific characteristics of bird species make them more or less sensitive to changes in habitat, thus determining their survival or extinction in fragmented landscapes (ALEIXO, 1999).

In addition, the results of studies in various ecosystems have shown that understanding just one or a few characteristics of the fragments is not enough to understand or predict how the avifauna will be affected (GIMENES & ANJOS, 2003). It is therefore essential to understand the various factors that interact with each other continuously and should always be considered together, such as the edge effect, size, shape, the adjacent matrix, as well as the degree of isolation of the fragments (ALEIXO, 2001; MMA/SBF, 2003).

The edge of a fragment or vegetation formation can be defined as the contact zone between a natural habitat and an anthropized one. In this way, the greater the proportion of the edge in relation to the area of a fragment, the smaller the central area, which is the effectively preserved area and the most similar to the original vegetation of the region (MMA/SBF, 2003).

The first response to the creation of an edge is the modification of the microclimate, which can affect the survival and reproduction of populations. At edges, soil and air humidity decrease, while soil and air temperature and light incidence increase, as do water vapor deficit and wind speed (MMA/SBF, 2003). The range of the edge effect is not a consensus in the literature and depends on various local factors, but it can be seen up to 400 meters from the edge of the forest to the interior (SILVA *et al.*, 2012).

The changes in the microclimate of the edge create conditions for the establishment of species that are different from those that were there before the edge was created. These conditions favor the establishment of species better adapted to the greater incidence of light, to the detriment of understory species (MMA/SBF, 2003). This can result in an edge bird community that is different from that inside the fragment, i.e. the effects that habitat fragmentation has on species are different (MMA/SBF, 2003).

The size of a habitat fragment can have a direct effect on the survival of bird populations. The smaller the fragment, the greater the influence of external factors on it. In small fragments, the dynamics of the ecosystem are probably determined by external rather than internal forces. Because they are smaller in area, small fragments are also home to small populations that are often unviable for maintaining the species (MMA/SBF, 2003).

Several studies have highlighted the greater intensity of edge effects in small fragments. The size of the fragment (and habitat) is an important factor in population dynamics and edge effects can further reduce the effective area of the fragment for certain species (MMA/SBF, 2003).

In addition to analyzing the size of each fragment, it is essential to analyze the shape of these fragments. The shape of a habitat fragment directly affects the ratio between the perimeter and the area of the fragment. The lower this ratio, the smaller the edge and the higher the ratio, the larger the edge (MMA/SBF, 2003).

Habitat fragments that are closer to a circular shape have a minimized edge-to-area ratio and, therefore, the center of the area is further away from the edges and, consequently, more protected from external factors. Areas that are more jagged (invaginated) have a higher proportion of edges than those that are less jagged (MMA/SBF, 2003) (Figure 3). The amount of a fragment's area represented by the edge is therefore a direct consequence of this relationship.

Figura 3. Diagram showing an example of a less jagged area (less invaginated) and a more jagged area (more invaginated), respectively. Source: MMA, 2003.

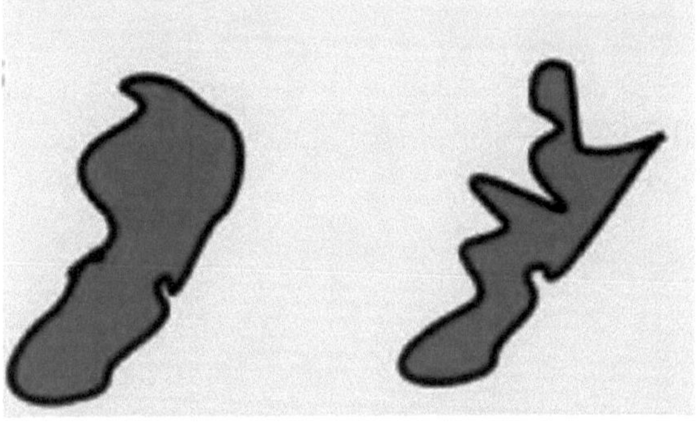

The term "matrix" refers to the mosaic of habitats around the fragment that have been modified by man, such as pastures, crops, agro-forestry systems (SAFs), plantations or regenerating forests, which surround the forest fragments (LAURANCE & VASCONCELOS, 2009). The matrix, as well as the distance between the fragments, determines the possibility of different species moving between the fragments. This mosaic of habitats around the fragments can represent a barrier for many bird species adapted to living inside forests, which prevents the flow of individuals between the fragments and can, over time, reduce the genetic variability of these populations. In fact, abrupt transitions between different types of vegetation represent a barrier to the dispersal of many birds, especially understorey birds (GIMENES & ANJOS, 2003; HAYES, 1995). According to Goosem (1997), even narrow linear clearings, opened up inside a forest to serve as roads, act as barriers for many species, especially those with low flight capacity and those accustomed to the interior of forests.

Therefore, the type of matrix surrounding the fragments influences the ability of organisms that existed in the original environment to remain in the remaining fragments. The more different the matrix is from the original environment, and the greater the distance between the fragments, i.e. the degree of isolation, the lower the chances of organisms remaining in the fragments. In this way, species that cannot cross the matrix to get from one fragment to another are more prone to local extinction. Several species of birds that live in forests are unable to cross altered environments (even though they can fly), i.e. they do not have flight autonomy. The microclimatic changes, especially the increase in luminosity and decrease in humidity, which follow the process of destruction and fragmentation, drive more sensitive species out of the understory. Other species, which naturally need large areas of original vegetation to survive, also lack the basic resources for survival in small remaining fragments, such as sufficient food and places to build nests, and also tend to disappear (MMA/SBF, 2003).

With regard to the degree of isolation, we can say that the distance between a fragment and other fragments or areas of forest can affect the movement of animals. Fragments close to other forest areas are more likely to receive immigrants than very isolated fragments. However, even a small deforested area can have a significant effect on the displacement of some species (LAURANCE & VASCONCELOS, 2009). Deforested strips only 15 to 100 m wide can act as barriers to the displacement of several species, including understorey bird species (LAURANCE *et al.* 2004).

1 OBJECTIVES

1.1 GENERAL OBJECTIVE

- To assess the conservation status of the fragments studied using the trophic guilds of the birds recorded as bioindicators.

1.2 SPECIFIC OBJECTIVES

- Correlate the results obtained by trophic guilds with the size and shape, edge effect, adjacent matrix and degree of isolation of the fragments.
- To collect data on the richness of birds in the areas studied, correlating this with the quality of the fragments studied.

2 MATERIAL AND METHOD

2.1 STUDY AREA

This study was carried out in four forest fragments, named F1 (20L 401828/9017131) with approximately 335 ha, F2 (20L 400516/9016010) with approximately 217 ha, F3 (20L 408805/9023889) with approximately 538 ha. and F4 (20L 394495/9035326) with approximately 308 ha. in the municipality of Porto Velho. Fragments F1, F2 and F3 are located on the right bank of the Madeira River and F4 is located on the left bank of the Madeira River (Figure 4). F1 and F2 are near highway BR 364, which connects Porto Velho/RO to Rio Branco/AC (Figure 5 and Figure 6, respectively) and F3, near highway BR 364, towards Cuiabà/MT (Figure 7). All three fragments are made up of vegetation characterized as dense submontane ombrophilous forest. F4 is located 5 km from the left bank of the Madeira River, close to the BR 319 highway, in the direction of Humaità/AM, at Fazenda Catarina (Figure 8). The vegetation of this forest fragment is characterized as lowland open ombrophilous forest with a predominance of palm trees.

Figure 4 - Landsat image with the boundaries of the four fragments, Porto Velho, Rondônia, Brazil. Source: www.googleearth.com. Accessed on April 10, 2013.

Figura 5. Landsat image showing the delimitation of fragment 1 (F1), Porto Velho, Rondônia, Brazil (see detail of the BR 364 highway to the right of F1). Source: www.googleearth.com. Accessed on April 10, 2013.

Figura 6. Landsat image with delimitation of fragment 2 (F2), Porto Velho, Rondônia, Brazil. Source: www.googleearth.com. Accessed on April 10, 2013.

Figura 7. Landsat image with delimitation of fragment 3 (F3), Porto Velho, Rondônia, Brazil. Source: www.googleearth.com. Accessed on April 10, 2013.

Figura 8. Landsat image with delimitation of fragment 4 (F4), Porto Velho, Rondônia, Brazil. Source: www.googleearth.com. Accessed on April 10, 2013.

2.2 METHODS

The study was carried out between January and February 2013. During this period, not only the areas inside the forest fragments themselves were sampled, but also their edges.

The method used was "observation points", which consists of observing the birds occurring at the point studied via auditory and visual recordings, with the observer remaining fixed at the point and collecting the data within a predetermined time interval (BIBBY *et al.*, 2000; RALPH *et al.*, 1995; RALPH & SCOTT, 1981).

The application of this method allowed a network of points to be established in the habitat studied, where the observer stayed for 20 minutes at each point in the morning and late afternoon, recording all the species seen and heard, as indicated by Alexandrino (2010).

One of the advantages of this method is that it is easy to standardize the number of sampling units, i.e. points, to be distributed within a heterogeneous environment (ALEXANDRINO, 2010).

Another advantage is that, according to Reynolds *et al.* (1980), an observer at a fixed point has more time to observe the birds and doesn't waste time looking at the ground as he moves, which is very important when dealing with tall, dense vegetation or rough terrain. Alexandrino (2010) also states that this method gives the observer a greater chance of visual and auditory contact with the bird species than a walking observer.

Studying wild animal populations is difficult because no survey method is suitable for all types of habitat and objectives, and each has its limitations (ANJOS, 2006; ANTUNES, 2005; BIBBY, 1992). Therefore, in order to complement this method, hikes were carried out along trails and accesses close to the edges.

The sampling effort lasted 14 days, totaling 112 effective hours of observation. Fifteen points were taken each day, restricted to the times used to walk near the edges. In this way, 210 20-minute samples were taken, with the sampling points considered independent, with a minimum spacing of 100 meters, randomly distributed along trails and accesses. The observations took place in the morning between 6:00-11:30 am and in the afternoon between 3:301-7:30 pm, as the birds are most active at these times, making it easier to detect them. A pair of Nikon 10x42 Monarch 3 All Terrain binoculars were used to observe the birds. Species not identified directly in the field were photographed, when possible, and their vocalizations recorded for later analysis, as proposed by Parker (1991).

Both the photographs and the recordings were taken with a Canon Powershot Sx10 digital camera. The recorded vocalizations were later compared with sound files available in the Xeno-Canto online database (www.xeno-canto.org; 2012), in order to correctly identify the species. The following sources were used to clarify the identification of certain species: Sick (1997); Sigrist (2008); Perlo (2009); Gwynne *et al.* (2010), as well as the help of experienced specialists.

The following criteria and acronyms were used to standardize field records: OBS: Observation (birds identified by simple observation); VOC: Vocalization (birds identified by vocalization); FOT: Photographed; GRA: Recording of vocalization (vocalization recorded for later identification and to document the record).

The nomenclature and taxonomic order used in this work are in accordance with those recommended by the Brazilian Committee of Ornithological Records (CBRO, 2011).

2.2.1 Trophic guilds and types of environments

The species were characterized by their trophic guilds and distribution in the environments, and their feeding habits were confirmed in specific literature (DARIO, 2009; DARIO, 2008; SICK, 1997; WILLIS, 1979). The following trophic guilds were considered: carnivore: feeding on large insects, small and large vertebrates; detritivore: feeding on dead animals; frugivore: feeding mainly on fruit; granivore: feeding on seed prey; insectivore: feeding mainly on insects that can be caught on the ground, in the air between vegetation and on tree bark; nectarivore: feeding mainly on nectar; omnivore: feeding on fruit, seeds, arthropods and small vertebrates; piscivore: feeding on fish (SCHERER, 2005). The guilds most sensitive to environmental disturbances will be considered throughout the discussion according to the most recent literature.

The following categories of environments were used: aquatic (streams, swamps and bodies of water), forest edge, tree canopy/understory, understory and pasture.

2.2.2 Factors affecting bird conservation in forest fragments

2.2.2.1 Edge effect

In this study, the edge effect will be considered to have an influence on the studied fragment with: little intensity, medium intensity or great intensity, following the recommendations of work carried out with bird groups in the Amazon region (SILVA *et al.*, 2012).

Following the method proposed in this study, using the Google Earth Pro software's perimeter marking and area calculation tool, the total area of the fragments was estimated, followed by the area that possibly suffers from the edge effect, which was only the portions of the fragments that were within 400 meters of any edge.

2.2.2.2 Size

Following the recommendations of Uezu (2006), the size of the fragments will be considered in this study: Small, when smaller than 100 ha, medium, when smaller than 250 ha and large those with an area greater than 250 ha.

1.1.1.1 Format

The following shape categories will be considered in this study: oval for the closest to a circular shape and elongated for the closest to a rectangular shape. In addition, the degree of invagination of the fragments will be subjectively categorized according to the following criteria: Not invaginated (those with regular edges), little invaginated (those with edges with few irregularities) and very invaginated (those with very irregular edges) according to the MMA (2003).

2.2.2.4 Matrix

The following types of matrix will be considered in this study: pasture, crops and urban areas (HAYES, 1995; GIMENES & ANJOS, 2003; LAURANCE & VASCONCELOS, 2009).

2.2.2.5 Degree of insulation

Following Laurance *et al.* (2004), this study will consider the following categories to determine the degree of isolation: Isolated or not isolated. The distance to other fragments will also be taken into account to determine whether they are: Juxtaposed (those with connections to other fragments); Close (fragments with distances of less than 15 m from each other); Distant (fragments with distances greater than or equal to 15 m) (LAURANCE *et al.* 2004).

3 RESULTS

Considering all the environments studied, 133 species of birds were recorded, distributed in 31 families and 19 orders. Families with the highest number of species were Thraupidae and Falconidae (7 species each), followed by Picidae (6 species) and Psittacidae, Tinamidae and Columbidae (5 species each) (Annex 1).

Looking at each fragment individually, in F1 there were 19 species, and in this area the families that stood out with the highest number of species were the Falconidae with three species, followed by the Picidae and Tinamidae families with two species.

In F2, 26 species were recorded, and in this area the family with the highest number of species was Picidae, followed by Ramphastidae, Bucconidae, Thraupidae, Columbidae, Charadriidae, Psittacidae and Tinamidae, each with two species recorded.

In F3, 42 species of birds were recorded. In this area, the Psittacidae family stood out, accounting for almost 12% of the species recorded.

In F4, 46 species were recorded. In this area, the family that stood out with the highest number of species was the Thraupidae, representing almost 15% of the total number of species recorded in this area.

The best represented trophic guilds were omnivores and insectivores, both with 24% of the total species recorded in all the fragments analyzed. Frugivores and carnivores were also well represented in all the fragments, with 21% and 18% respectively of the total number of species recorded. On the other hand, detritivores, granivores, nectarivores and piscivores were the least represented (Graph 1).

Graph 1. Average percentage representation of the different trophic guilds recorded in the four fragments studied.

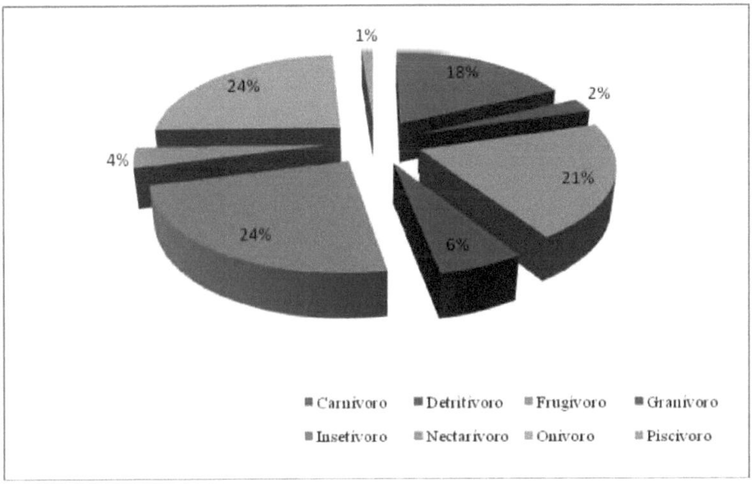

Analyzing the trophic guilds of the fragments separately, in F1 we have insectivores and frugivores as the most representative guilds in this area and, of the total species, more than 60% make up these two guilds. The carnivore guild also stood out, with 21% of the species recorded in this fragment.

Graph 2. Average percentage representation of the different trophic guilds recorded in F1.

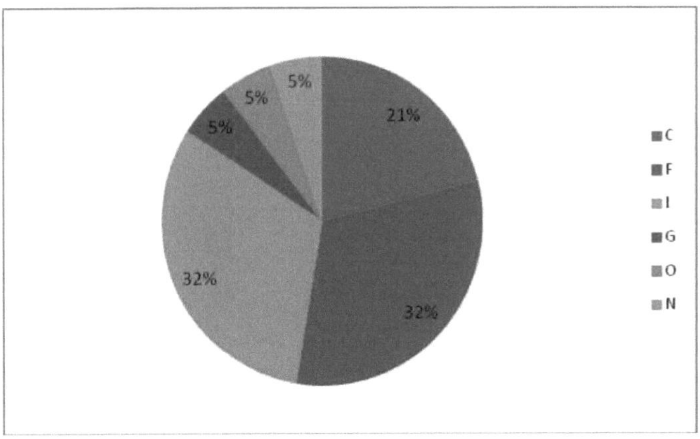

In F2, the most representative guilds were omnivores and insectivores, both with 27%, followed by frugivores with 23% of the species recorded in this area.

Graph 3. Average percentage representation of the different trophic guilds recorded in F2.

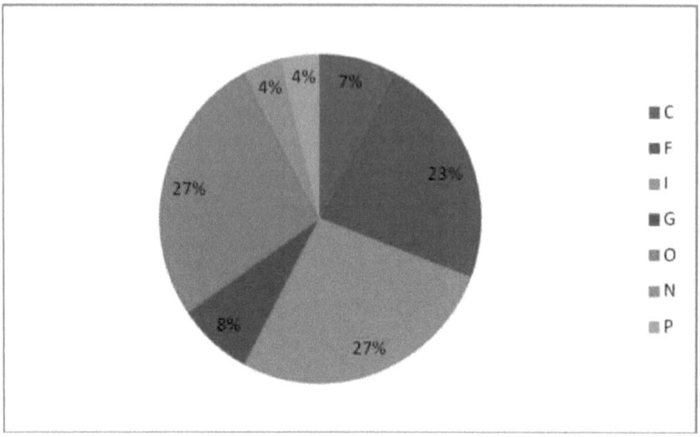

In the analysis of F3, insectivores are the most representative guild, with 31% of the total species in this fragment, followed by frugivores, with 24% of the species.

Graph 4. Average percentage representation of the different trophic guilds recorded in F3.

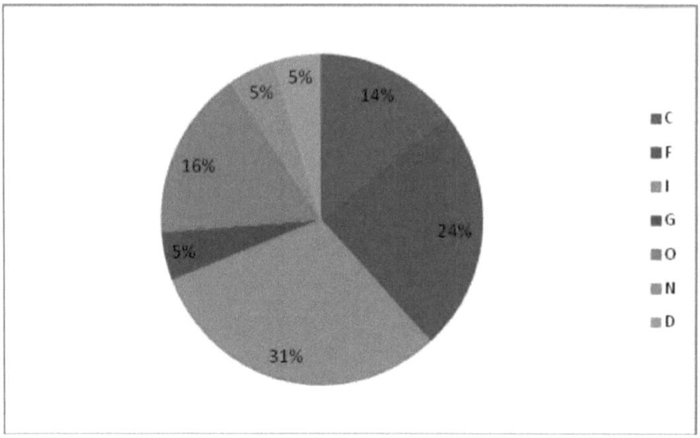

In F4, omnivores stood out, accounting for 28% of all species in this area. The insectivore guild also stood out, with 26%.

Graph 5. Average percentage representation of the different trophic guilds recorded in F4.

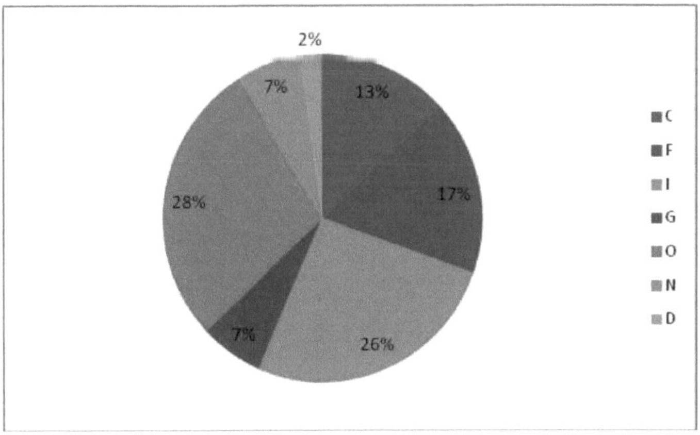

3.1 Factors affecting bird conservation in forest fragments

In the analysis of the factors that together affect the quality of the environment, such as the size, shape, adjacent matrix and degree of isolation of the fragments, the following results were observed:

3.1.1 Edge effect

When analyzing F1, we observed that this environment probably suffers a great deal from the influence of the edge effect, as only 10.69 ha of the 335 ha can probably be considered a properly preserved area.

In the analysis of F2, it was observed that the entire fragment is probably suffering from the edge effect, a factor made worse by the transmission line built as part of the construction of the hydroelectric dams, tracing an opening approximately 15 meters wide, thus increasing the contact zone between a natural habitat and an anthropized one.

In F3, the analysis showed that this fraction probably suffers a great deal from the influence of the edge effect. Of the 538 ha, only 23.50 ha can probably be considered a properly preserved area, i.e. an area with vegetation more similar to the original.

In the analysis of F4, it was observed that, as in F2, the entire fragment is probably suffering from the edge effect. In this case, it is likely that the aggravating factor is due to the various clearings within the forest caused by timber and gravel extraction.

3.1.2 Size of fragments

The size of the fragment is another important factor for the survival of bird populations, and the smaller the fragment, the greater the influence of external factors, such as the edge effect, on it (MMA, 2003).

In addition, the mere fact of habitat reduction also leads to population declines, as some species need large areas to survive. A classic example is the *harpy* eagle, *Harpia harpyja*, which is estimated to need an area of 100 to 200 km^2 of preserved forest to survive (LAURENCE & VASCONCELOS, 2009; SICK, 1997).

F1, F3 and F4 were considered large fragments, as their areas are larger than 250 ha. Only F2 was considered medium-sized, as its area is less than 250 ha. In this study, no fragment was considered small, as none of them had an area smaller than 100 ha.

3.1.3 Format of the fragments

When analyzing the shape of the fragments, we found that F1 and F3 are closer to rectangular in shape, which is not considered ideal, as the centers of the areas are closer to the edges and, consequently, less protected from external factors (Figure 5 and Figure 7, respectively). F2 and F4, on the other hand, have shapes closer to the oval shape, which is considered the recommended shape because, as opposed to the shape closer to the rectangular shape, fragments with this shape have the center of the area further away from the edges and, consequently, are better protected from external factors, such as the edge effect (Figure 6 and Figure 8, respectively).

3.1.4 Distance between fragments and degree of isolation

When analyzing the distance between fragments and the degree of isolation, F1 and F2 were considered to be distant in relation to other closer fragments, and were therefore also classified as isolated, since according to Laurance *et al.* (2004) deforested strips only 15 to 100 m wide can act as barriers to the movement of various bird species, and both fragments are surrounded by openings over 15 m wide. F3 and F4, on the other hand, are connected to other fragments by strips of native vegetation, which vary in width from 754 to 2,454 meters, i.e. they are not isolated, allowing the flow of individuals between the fragments, probably preventing a decrease in the genetic variability of these populations.

3.1.5 Adjacent matrix

In this analysis, it was observed that the type of matrix adjacent to all the fragments was pasture, so because this type of matrix is very different from the original environment of the fragments, it is possible that many species are unable to cross this matrix to go from one fragment to another, thus making them more prone to local extinction.

Table 3. Factors that may affect the conservation of avifauna in the forest fragments analyzed in this study. Edge effect: Intense (INT); Size: Large (G) and Medium (M); Shape: Elongated (ALON) and Oval (OVAL); Degree of edge invagination: Very invaginated (M.I.); Not invaginated (N.I.); Slightly invaginated (P.I.); Distance between fragments: Distant (D) and Juxtaposed (J); Degree of isolation: Isolated (I); Not isolated (N.I.); Adjacent matrix: Pasture (P).

	Edge effect	Size	Format	Degree of edge invagination	Distance between fragments	Degree of insulation	Adjacent matrix
F1	INT	G	ALON	M.I.	D	I	P
F2	INT	M	OVAL	N.I.	D	I	P
F3	INT	G	ALON	M.I.	J	N.I.	P
F4	INT	G	OVAL	P.I	J	N.I.	P

4 DISCUSSION

When analyzing the fragments separately, the F1 avifauna showed that frugivores and insectivores had the highest representation, both with 32%, followed by the carnivore guild with 21% of the total species recorded in this area. The representation of frugivores, especially the large species (e.g. *Trogon viridis* and *Pteroglossus castanotis)*, in this study area is an important factor, as the species in this guild favor the maintenance of forest fragments, being responsible for the dispersal of 39% to 77% of the plant species in a tropical forest, contributing to its conservation (STILES, 1985). In particular, *Penelope superciliaris* (Jacupemba) also belongs to the frugivore guild, a bird that forages on the ground and in the understory. This species is considered to be more sensitive to forest fragmentation due to its low dispersal capacity and need for large foraging areas. Thus, its presence in this fragment indicates that this area still has good environmental quality (DONATELLI *et al.* 2007; STOUFFER & BIERREGAARD 1995). However, the presence of this species in highly fragmented areas makes them more susceptible to hunting pressure or even predation, tending towards local extinction (PERES, 2001).

With regard to insects, which were also well represented in this fragment, despite being considered by some authors (DARIO, 2010; SICK, 1997) as generalist species, due to the fact that they have a high capacity to adapt to altered environments, i.e. they are undemanding in terms of environmental resources, we should highlight the presence of the Picidae family represented in this area by the woodpeckers (*Picumnus aurifrons* and *Melanerpes cruentatus*), Despite being part of this guild, insectivorous birds that climb trunks and branches are considered to be very sensitive species to fragmentation, quickly becoming extinct as their habitat diminishes (WILLIS, 1979). Therefore, the presence of these species in the fragment studied indicates that this area still has good environmental quality.

The presence of carnivores in this fragment (e.g. *Herpetotheres cachinnans, Falco femoralis, Daptrius ater*) indicates the presence of relatively balanced ecosystems of great biological value (WILLIS 1979), since carnivore species are very important to the environment in which they reside, are predators usually at the top of the trophic chain, also needing large areas to forage and are harmed by degradation and by the change in the abundance of their prey (ALEIXO, 1999).

When comparing the results obtained through the analysis of the trophic guilds with the physical characteristics of the fragment, it was found that these were not in agreement, since, contrary to the results of the guilds, the physical characteristics indicate that the fragment is probably not of good quality.

The guilds recorded in F1 are considered good indicators of environmental quality due to the sensitivity of these species to changes in the environment. However, this sensitivity makes them vulnerable to significant population reductions or even local extinction. This tendency is due to the fact that there are several factors that interact with each other on an ongoing basis, which can directly affect the conservation of the avifauna in these fragments, one of which is the edge effect, a factor that can be aggravated in synergy with other factors, such as the shape, which in this fragment is not ideally presented. In this study, it was observed that 96.81% of this area is probably being greatly influenced by the edge effect, if we consider Silva *et al.*'s (2012) criterion of 400 meters of edge effect. The changes created in the microclimate of this area due to this factor favor the establishment of species that are less demanding in relation to environmental resources, to the detriment of those that are more demanding, which could result in a bird community that is very different from the one previously present in the fragment.

The distance between fragments, and the consequent degree of isolation, are other factors that may contribute to the decline of local populations of species that are sensitive to environmental changes in this area. When analyzing the distance between fragments and the degree of isolation, F1 was categorized as distant and therefore isolated, as well as having an adjacent matrix made up of pastures. These factors prevent the movement of several species recorded in this area, such as those that do not cross (or rarely cross) open areas due to their low flight capacity and/or behavioral or physiological limitations (e.g. *Penelope superciliaris* and *Trogon viridis*), thus making them more prone to local extinction.

In the analysis of F2, we observed that 54% of the species recorded belonged to the omnivore and insectivore guilds (27% each), known as synanthropic, i.e. species that expand their geographical distribution as the original vegetation is suppressed (DARIO, 2010). In this area, the frugivore guild also stood out, accounting for 23% of the species recorded.

Omnivores represent species that are able to adapt to different diets in different environmental conditions, which makes them increasingly abundant in fragmented environments (MOTTA-JÙNIOR, 1990; WILLIS, 1979). According to Sick (1997), these species have a high degree of plasticity with regard to the impacts caused by human activities in altered landscapes and a high capacity to adapt to altered environments, i.e. they are undemanding with regard to environmental resources. In this way, the high representativeness of this guild (*Saltator maximus, Ramphocelus carbo, Cacicus cela, Tyrannus melancholicus, Leptotila rufaxilla, Vanellus chilensis*) in the area studied may indicate a reduction in habitat quality, probably due to the high level of anthropogenic activity in the area.

The insectivores, as in F1, were mainly represented by the Picidae (*Melanerpes candidus, Melanerpes cruentatus* and *Veniliornis affinis*). The guild of "branch and trunk insectivores" is made up of species whose diet consists of insect larvae found in dead trees that are still standing. According to Sick (1997), these birds are considered specialists and cannot be generalized when it comes to the feeding habits of this group. An environment that has several representatives of the species in this group indicates a good quality forest environment. However, it is possible that the greater presence of "branch and trunk insectivore" species in this fragment (as opposed to frugivores, which is probably linked to the good mobility that some species in this group have) is due to the greater availability of food resources for species in this group, which does not necessarily indicate that this fragment is of good quality.

When comparing the results obtained through the trophic guilds with the data obtained for the physical characteristics of F2, it was observed that these are largely in agreement, as they probably indicate the low quality of this environment. Most of the species belonging to the guilds recorded in this fragment are considered generalists, so these species benefit from anthropotized environments. However, in the insectivore guild, the Picidae family stood out as an indicator of good habitat quality. However, it is believed that the detection of these species in this fragment is due exclusively to the presence of several dead trees that are still standing, which provide a great availability of food resources for this group in this area. With regard to the frugivores recorded in this area, it is believed that this is due to the fact that, although this fragment is surrounded by pasture, there are still some trees that may be serving as a "springboard" for these species to move from one fragment to another, since species such as *Ramphastos tucanus* and *Pteroglossus castanotis* are canopy birds that have good mobility and can move more easily through the modified landscape and thus persist in the fragments. Therefore, the presence of the Picidae family and the Ramphastidae does not necessarily indicate that the fragment is of good quality.

As a result of the influence of these factors, it was found that although this fragment is of an ideal size and shape, it is heavily influenced by the edge effect which, according to this study, is probably affecting 100% of this area. In addition, when analyzing the degree of isolation, this fragment was considered to be distant in relation to other nearby fragments, and was therefore also classified as isolated, a factor made worse by the type of adjacent matrix, which in this case consists of pasture.

The lack of forest cover around these fragments can represent an important barrier for several species, such as *Crypturellus cinereus* and *Crypturellus parvirostris*, birds with low flight capacity recorded in this fragment. These factors, which are negatively influencing the quality of this fragment, could, over time, cause a further decline in these species, which respond negatively to changes in the environment and could be driven to extinction, which would probably only be observed in long-term studies.

In F3, the guild that stood out the most was insectivores (31%), represented mainly by species from the Galbulidae and Dendrocolaptidae families. Species from the Galbulidae family are considered undemanding in terms of environmental resources. On the other hand, species from the Dendrocolaptidae family, as well as those from the Picidae family, are part of the group of "trunk-climbing birds" and the presence of these species in the study site, with not many dead trees still standing in this area, may indicate that these birds were not only present due to the great availability of food resources. Therefore, the presence of these species in this area may indicate that the environment is of good quality, because these species are extremely closely related to forest environments, and are the ones that disappear most quickly after disturbances (SOARES & ANJOS, 1999; STOUFFER & BIERREGAARD, 1995; WILLIS, 1979).

After insectivores, the frugivore guild also stood out with 24% of the total species recorded in this fragment. As described above, frugivores are indicators of a good quality forest environment.

When comparing the results obtained through the trophic guild analysis and the data related to the physical characteristics of F3, it was found that they do not agree, since the Dendrocolaptidae family was highlighted in this fragment among the insectivore guild, due to the fact that it is an indicator of good environmental quality, since the species of this family are not present in this area exclusively because of the richness of food, as has already been discussed.

The guild of frugivores is mainly represented by the Psittacidae and Ramphastidae families, which would not necessarily indicate that this environment is of good quality. However, species such as *Trogon viridis, Trogon curucui* and *Querula purpurata*, which were also recorded in this area, are good indicators of environmental quality, due to the fact that these species are associated with well-preserved environments.

Although several species were recorded, indicating that this fragment can still be seen as having good environmental quality, factors that affect the conservation of avifauna in forest fragments can change this picture. In this study, it was observed that 95.63% of the fragment may be suffering from the influence of the edge effect, which leads to various factors that diminish the quality of this environment, including an increase in luninosity that negatively affects species adapted to the shaded interior of the forest. In addition, although this fragment is considered large, it is not ideally shaped, which is one of the main aggravating factors of the edge effect. Another important factor is that this fragment, despite being in the middle of a matrix formed by pasture, is connected to other fragments by ecological corridors, which allows the flow of individuals between the fragments, probably preventing a decrease in the genetic variability of these populations.

In F4, the guild that stood out the most was the omnivores (e.g. *Saltator maximus, Ramphocelus carbo, Tangara episcopus, Tangara palmarum* and *Cacicus cela*) as already mentioned in the discussion of the results found in F1. They are considered by several authors to be generalist species (MOTTA-JÛNIOR, 1990; SICK, 1997; WILLIS, 1979), due to their high capacity to adapt to altered environments, making them undemanding in terms of environmental resources. The high number of species recorded in this fragment belonging to this guild is probably due to the fact that this environment has been subjected to a great deal of anthropogenic activity and to the opening of several artificial clearings within it.

Insectivores also featured prominently in this fragment, where the guild is represented by some species considered to be undemanding in terms of environmental resources (e.g. *Progne tapera, Piaya cayana, Crotophaga ani* and *Nyctiphrynus ocellatus*), according to various authors (MOTTA-JÛNIOR, 1990; SICK, 1997; WILLIS, 1979). On the other hand, "trunk-climbing insectivores" were recorded in this fragment, represented by the Picidae and Dendrocolaptidae families. These groups are considered to be very sensitive to habitat fragmentation and their presence indicates good habitat quality (WILLIS, 1979). As has already been discussed, this fragment suffers great anthropogenic pressures, but the presence of this group may indicate that this area still has the necessary resources for these species.

When we compare the results obtained through the trophic guild analysis, we see that most of them are consistent with the physical characteristics presented in F4, since the guilds recorded in this fragment were those that are undemanding in terms of environmental resources, making it easy for them to adapt to areas that have undergone environmental changes. However, species from the Picidae and Dendrocolaptidae families were recorded, which are considered to be sensitive to environmental changes. It is likely that factors affecting the conservation of avifauna in forest fragments will cause these more sensitive species to decline or even become locally extinct. In this study it was observed that both the size and shape of this fragment are as recommended, but it is likely that the whole fragment is suffering from the edge effect, a factor that may be aggravated by the various clearings within it, caused by the extraction of wood and gravel. It is worth noting that this fragment is in the middle of a matrix formed by pasture, but it is connected to other areas by ecological corridors, i.e. it is not isolated, a factor that contributes to the variability of the genes of the populations present in this habitat, perhaps preventing a decline in species.

CONCLUSION

This study showed that possibly none of the fragments studied had good environmental quality, since even the fragments where the most representative guilds were those of species typical of understory environments, which indicate good environmental quality, or that most of the physical characteristics of the fragments were in some way suitable for good habitat quality, there was never total corroboration between these factors.

This suggests that, possibly, in the fragments where the most representative guilds indicate good habitat quality, they are or may be declining due to various factors that reduce environmental quality, such as the large edge effect, large distances between other fragments, among others. In the long term, there may be a reduction in the proportion of these guilds to the detriment of an increase in the proportion of guilds that indicate poor quality, such as omnivores.

It is therefore necessary to review the way in which forest fragments are distributed and formed so that they can better adapt to the physical characteristics considered to be favorable to habitat conservation, for example, large size, little invagination and matrices closer to the original vegetation as ecological corridors. Possibly, if most of the forest fragments near the city of Porto Velho had these characteristics, they would be more suitable for conserving local birdlife.

REFERENCES

ALEIXO, A. Effects of selective logging on a bird community in the Brazilian Atlantic Forest. ***Condor***, 101: p. 537:548. 1999.

ALEIXO, A. Conservaçao da avifauna da Floresta Atlàntica: efeitos da fragmentaçao e a importância de florestas secundàrias, p. 199-206. In: J. L. B. Albuquerque, J. F. Cândido Jr., F. C. Straube ***and*** A. L. Roos (eds.) ***Ornithology and Conservation - From Science to Strategies***. Tubarao: Unisul. 2001.

ALEIXO, A; VIELLIARD, J.M.E. Composiçao e dinâmica da avifauna da mata de Santa Genebra, Campinas, Sao Paulo, Brazil. ***Revista brasileira de zoologia***, v.12, n.3, p.493- 511. 1995.

ALMEIDA, A. F. Monitoring fauna and their habitats in forested areas. In: ***Technical Series***. IPEF, v. 12, n. 31, p. 85-92. 1998.

AMBIENTE BRASIL. Amazônia, a lot of biodiversity and little knowledge. Available at: <http://ambientes.ambientebrasil.com.br>. Accessed on: January 2013. 2013

ANDRADE, F.T de; FISCH, S.T.V.; FORTES NETO, P.; BATISTA, G.T. Avifauna in fragmented tropical forests: indicators of sustainability in Hydroelectric Power Plants. ***Repositório Eletrônico Ciências Agràrias, Coleçao Ciências Ambientais***, http://www.agro.unitau.br/dspace. p. 1-11. 2011.

ANGELO, H.; SA, S. P. P. Deforestation in the Brazilian Amazon. ***Ciência Florestal***, Santa Maria, v. 17, n. 3, p. 217-227, jul.-set, 2007.

ANJOS, L. Bird communities in five Atlantic forest fragments in Southern Brazil. ***Orn. Neotr***. 12:11-27, 2001; BIERREGAARD, R. O. J.; STOUFER, P. C. Understory birds and dynamic habitat mosaics in Amazonian rainforests. In: LAURENCE, W.F.; BIERREGAARD, R.O.J. (Org.). Tropical forest remnants: ecology, management, and conservation of fragmented communities. Chicago: University of Chicago Press, 1997. p.138-155. 2001.

ANJOS, L. Bird species sensitivity in a fragmented landscape of the Atlantic Forest Southern Brazil. ***Biotropica***. 32 (2): 229-234. 2006.

ANTUNES, A.Z. Changes in bird community composition over time in a forest fragment in southeastern Brazil. ***Ararajuba*** 13 (1):47-61. 2005.

AVISSAR, R., PIELKE, R.A. A parameterization of heterogeneus land surfaces for atmospheric numerical models ans its impact on regional meteorology. ***Monthly Weather Review***, 117, 21132136. 1989.

BIBBY, C.J.; BURGESS, N.D.; HILL, D.A. Bird census techniques. ***London, Academic Press***. 258p. 1992.

BIERREGAARD, R.O. & GASCON, C. The Biological Dynamics of Forest Fragments Project: Overview and History of a Long-Term Conservation Project. Pp. 31-45 *In*: R.O. Bierregaard, C. Gascon, T.E. Lovejoy & R. Mesquita (eds). Lessons From Amazonia: The Ecology and Conservation of a Fragmented Forest. ***Yale University Press, New Haven***. 478p. 2001.

BRAZILIAN ORNITHOLOGICAL REGISTRY COMMITTEE, 2011. Lists of the birds of Brazil. 10ª Ediçao, 25/1/2011, Available at <http://www.cbro.org.br>. Accessed on: February 2013.

CORLETT, R.T. Environmental heterogeneity and species survival in degraded tropical landscapes. In: M.J. Hutchings, E.A. John & A.J.A. Stewart (eds.). The ecological consequences of environmental heterogeneity. pp. 333-355. ***British Ecological Society***, London. 2000.

DARIO, F. R. Trófica estrutura da avifauna em fragmentos florestais na Amazônia Oriental. ***ConScientiae Saùde***, v. 7, n. 2, p. 169-179. 2008.

DARIO, F. R. Avifauna of Atlantic Forest Fragments in Southern Espírito Santo. ***Biotemas***. 23 (3): p. 105-115. 2010.

DONATELLI, R.J; FERREIRA, C.D.; DALBETO, A.C. & POSSO, S.R. Comparative analysis of the bird assemblage in two forest remnants in the interior of the State of Sao Paulo, Brazil. ***Revista Brasileira de Zoologia***, 24: 362-375. 2007.

FEARNSIDE, P.M. Deforestation in the Brazilian Amazon: history, rates and consequences. ***Megadiversity***, v.1, n.1, p113-122. 2005.

FEARNSIDE, P. M. Deforestation in Amazonia: Dynamics, Impacts and Control. ***Acta Amazônica***. VOL. 36(3): 395 - 400. 2006.

FERREIRA, G. L. F. Plan for prevention, control and sustainable alternatives to deforestation in Rondônia, 2009 - 2015. ***State Secretariat for Environmental Development - SEDAM.*** 2009.

GASCON, C.; LAURENCE, W.F. & LOVEJOY, T.E. Forest fragmentation and biodiversity in Central Amazonia. In: ***Conservation of biodiversity in tropical ecosystems***. Garay, I & Dias, B. (eds.), Editora Vozes, p: 174-189. 2001.

GIAMBELLUCA, T.W.; ZIEGLER, A.D.; NULLET, M.A.; TRUONG, DM; TRAN, LT. Transpiration in a small tropical forest patch. ***Agricultural And Forest Meteorology*** 117 (1-2), 1-2.

2003.

GIMENES, M. R. & ANJOS, L. dos. Effects of forest fragmentation on bird communities. Acta Scientiarum. Biological Sciences, 25(2): 391-402. Goerck, J. M. 1997. Patterns of rarity in the birds of the Atlantic forest region of Brazil. ***Conservation Biology***, 11: 112-118. 2003.

GOOGLE. ***Google Earth***. Developed by Google Inc.: Mountain View. 2011. Available at: < http://earth.google.com>. Accessed on: January 2013.

GOOSEM, M. Internal fragmentation: the effects of roads, highways, and powerline clearings on movements and mortality of rainforest vertebrates. Chicago: ***The University of Chicago Press***. Chap. 16, p. 241-255. 1997.

GWYNNE J.A.; MARTHA, A.; RIDGELY, R.S.; TUDOR, G. ***Birds of Brazil. Pantanal & Cerrado***. Ed. Horizonte. 336p. 2010.

HAYES, F.E. Status, distribution and biogeography of the birds of Paraguay. Loma Linda: Loma ***Linda University***. 1995.

HIGGS, A.J.; USHER, M.B. Should nature reserves be large or small? ***Nature,*** London, v. 285, p. 568-569. 1980.

ST, A. Biogeography and ecology of forest bird communities. The Hague: ***SPB Academic Publishing***,. ch. 27, p. 387-393. 1990a.

HOLMES, R.T. The structure of a temperate deciduous forest bird communities: variability in time and space. *In*: JAMES, F.C. Relationships between temperate forest bird communities and vegetation structure. ***Ecology***, Washington, D.C., v. 63, n. 1, p. 159-171, 1982. 1990b.

IBGE - Brazilian Institute of Geography and Statistics (IBGE), 2013. Available at: <http://www.ibge.gov.br>. Accessed on: February 2013.

IMAZON - Instituto do Homem e Meio Ambiente da Amazônia. Deforestation and Forest Degradation in the Amazon Biome (2000 - 2010), p. 2. Belém, Parà, Brazil. 2011.

IMAZON - Instituto do Homem e Meio Ambiente da Amazônia. Boletim Transparência Florestal da Amazônia Legal, Belém, PA - Brasil, p. 12. 2013.

IMAZON - Instituto do Homem e Meio Ambiente da Amazônia. Map produced at the Imazon Geotechnology Center (CGI). Belém, Parà, Brazil. 2013.

INPE - National Institute for Space Research. Sao José dos Campos, SP, 2007. Available at: <http://www.obt.inpe.br/prodes>. Accessed on: February 2013

INPE - National Institute for Space Research. Sao José dos Campos, SP, 2012. Available at: <http://www.obt.inpe.br/prodes>. Accessed on: February 2013.

KAPOS, V. Efects of isolation on the water status of forest patches in the Brazilian Amazon. ***Journal of Tropical Ecology***, 5, Part-2, 173-185. 1989

LAPS, R.R. Effect of fragmentation and habitat alteration on the avifauna of the Uma Biological Reserve, Bahia. PhD Thesis, State University of Campinas, Sao Paulo. 177 pp. 2006.

LAURANCE, W.F. & BIERREGAARD R.O. JR. "Preface: A crisis in the making". In: Tropical Forest Remnants: Ecology, Management and Conservation of Fragmented Communities. Laurance, W.F. and Bierregaard R.O. Jr. (eds). ***University of Chicago Press***, Chicago, USA. 1997

LAURANCE, S.G.; STOUFFER, P.C. & LAURANCE,W.F. Effects of road clearings on movement patterns of understory rainforest birds in central Amazonia. ***Conservation Biology***, 18: 1099-1109. 2004.

LAURANCE, W.F.; & VASCONCELOS; H.L. Consequencias ecológicas da fragmentaçao forest in Amazonia. ***Oecologia Brasiliensis***, 13(3): 434-451. 2009.

MACARTHUR, R.H.; WILSON, E.O. The theory of island biogeography. Princeton: ***Princeton University Press***. 203 p. 1967.

MARGULIS, S. Causas do desmatamento da Amazônia brasileira. 1ª ed. Brasilia: World Bank, 100 p. 2003.

MINISTRY OF THE ENVIRONMENT - MMA. Ecosystem Fragmentation: Causes, effects on biodiversity and public policy recommendations. Secretariat of Biodiversity and Forests (SBF), Brasilia, 510 p. 2003.

MOTTA-JUNIOR, J. C. Trophic structure and composition of the avifauna of three terrestrial habitats in the central region of the state of São Paulo. ***Ararajuba***, *1*: 65-71. 1990.

PARKER, T.A. On the use of tape recorders in avifaunal surveys. ***Auk***, v.108, p. 443-444. 1991.

PERES, C.A. Synergistic effects of subsistence hunting and habitat fragmentation on Amazonian forest vertebrates. ***Conservation Biology***, 15: 1490-1505. 2001.

PÉRICO, E.; CEMIN, G.; LIMA, D.F.B.; REMPEL C. Effects of habitat fragmentation on animal communities: using geographic information systems and landscape metrics to select suitable areas for testing. ***Anais XII Simpòsio Brasileiro de Sensoriamento Remoto***, Goiânia, Brasil, 16-21, INPE, p. 2339-2346. 2005.

PERLO, B.V. A Field guide to the birds of Brazil. Oxford USA Trade. 465p. 2009.

POULIN, B.; LEFEBVRE, G. and McNEIL, R. Characteristics of feeding guilds and variation in diets of bird species of three adjacent tropical sites. ***Biotropica*** 26:187-197. 1994.

REYNOLDS, R. T. *et al.* A variable circular-plot method for estimating bird numbers. ***The Condor***, Berkeley, v. 82, p. 309-313. 1980.

ROOT, R.B. The niche exploitation pattern of the bluegray gnatcatcher. ***Ecological monographs,*** v.37, n.l, p.317350. 1967.

ROBBINS, C.S. Census techniques for forest birds. In :Workshop Management of Southern Forests for Nongame Birds, 1978, Atlanta. Asheville: USDA Forest Service, p. 142-163. (General Technical Report, 14). 1978.

SCHERER, A.; SCHERER, S. B.; BUGONI, L.; MOHR, L.V.; EFE, M.A.; HARTZ, S.M. Estrutura trófica da Avifauna em oito parques da cidade de Porto Alegre, Rio Grande do Sul, Brasil, ***Ornithologia*** 1(1):25-32. 2005.

SIMBERLOFF, D. & DAYAN, T. The guild concept and the structure of ecological communities. ***Annual Ver. Ecol. Syst***. 22:115-143. 1991.

SICK, H. ***Ornitologia Brasileira***. Rio de Janeiro: Editora *Nova Fronteira*, 912 p. 1997.

SIGRIST, T. ***Birds of the Brazilian Amazon***. Vinhedo - SP. Publisher: *Avis Brasilis, 471* p. 2008.

SILVA, J. V. C.; CONCEIÇÂO, B. S.; ANCIÂES, M. Use of secondary forests by understorey birds in a fragmented landscape in central Amazonia. ***Acta Amazonica,*** vol. 42(1) p. 73 - 80. 2012.

SKOLE, D; TUCKER, C. Tropical Deforestation And Habitat Fragmentation In The Amazon - Satellite Data From 1978 To 1988 (Vol 260, Pg 1909, 1993). ***Science*** 261 (5125): 1104-1104. 1993.

SOARES, E. S. & ANJOS, L. dos. Effect of forest fragmentation on trunk and branch climbing birds in the Londrina region, northern Paranà state, Brazil. ***Ornitologia Neotropical***, 10: 61-68. 1999.

STOUFFER, P.C. & R. O. BIERREGAARD JR. 1995. Use of Amazonian forest fragments by understory insectivorous birds: effects of fragment size, surrounding secondary vegetation, and time since isolation. ***Ecology*** 76:2429-2445.

STILES, F.G. On the role of birds in the dynamics of neotropical forests. Pp. 49-59. *In*: Diamond, A. W. & Lovejoy, T. E. (eds.). ***Conservation of tropical forest birds***. ICBP Technical Publications. 1985.

TABARELLI, M.; DASILVA, M.J.C. & GASCON, C. Forest fragmentation, synergisms and the impoverishment of neotropical forests. ***Biodiversity and Conservation***, 13: 1419-1425. 2004.

TERBORGH, J. *et al.* Transitory states in relaxing ecosystems of land bridge islands. In: LAURANCE, W.F.; BIERREGAARD, R.O. (Ed.) Tropical forest remnants: ecology, management and conservation of fragmented communities. Chicago: ***The University of Chicago Press***, ch. 17, p. 256-274. 1997.

UEZU, A. Composiçao e estrutura da comunidade de aves na paisagem fragmentada do Pontal do Paranapanema, SP. 193 p. PhD thesis, Institute of Biosciences, University of Sao Paulo. 2006.

VIANA, V.M. *et al.* Dynamics and restoration of forest fragments in the Brazilian Atlantic Moist Forest. *In*: LAURANCE, W.F.; BIERREGAARD, R.O. (Ed.) Tropical forest remnants: ecology, management and conservation of fragmented communities. Chicago: ***The University of Chicago Press***. ch. 23, p. 351-365. 1997.

VIEIRA, L.J.S.; COSTA, S.S. Melo. da; OLIVEIRA, C.H. de; LOPES, M.R.M.; SILVEIRA, M. 2006. São Francisco stream basin - Rio Branco (AC): characterization and anthropic impacts. In: Marco Antônio Oliveira. (Org.). Pesquisa socioparticipativa na Amazônia Ocidental. 1 ed. Rio Branco (AC): EDUFAC - ***Editora da Universidade Federal do Acre,*** v. 1, p. 216-234.

XENO-CANTO. Xeno-Canto; Foundation receives generous financial support from Naturalis Biodiversity Center. 2013. Available at: <*http:*//www.xeno-canto.org>. Accessed February 2013.

WARBURTON, N.H. Structure and conservation of forest avifauna in isolated rainforest remnants in tropical Australia. *In*: LAURANCE, W.F.; BIERREGAARD, R.O. (Ed.) Tropical forest remnants: ecology, management and conservation of fragmented communities. Chicago: ***The University of Chicago Press.*** ch. 13, p. 190-206. 1997.

WILCOVE, D.S.; ROBINSON, S.K. 1990. The impact of forest fragmentation on bird communities in Eastern North America. *In*: KEAST, A. Biogeography and ecology of forest bird communities. (Ed.) ***The Hague: SPB Academic Publishing***, chap. 21, p. 319-331.

WILLIS, E. O. The composition of avian communities in remnant woodlots in southern Brazil. ***Papéis Avulsos de Zoologia***, *33*: 1-25. 1979.

APPENDIX I

List of bird species recorded in the environments studied. Trophic guilds: (C) carnivore, (D) detritivore, (F) frugivore, (G) granivore, (I) insectivore, (N) nectarivore, (O) omnivore, (P) piscivore. Habitats (main environments of occurrence in this study): (A) aquatic, (B) forest edge, (C) tree canopy and understorey, (P) pasture, (S) understorey. Field record: OBS: Observation (birds identified by simple observation); VOC: Vocalization (birds identified by vocalization); FOT: Photographed; GRA: Recording of vocalization (vocalization recorded for later identification and to document the record). F1 (20L 401828/9017131); Fragment located on the right bank of the Madeira River, near the BR 364 highway, which connects Porto Velho/RO to Rio Branco/AC; F2 (20L 400516/9016010); Fragment located on the right bank of the Madeira River, near the BR 364 highway, which connects Porto Velho/RO to Rio Branco/AC; F3 (20L 408805/9023889); Fragment located on the right bank of the Madeira River, near the BR 364 highway, towards Cuiaba/MT; F4 (20L 394495/9035326); Fragment located 5 km from the left bank of the Madeira River, near the BR 319 highway, towards Humaita/AM, at Fazenda Catarina.

Name of Τάχοη	Popular Name	Guild trophic	Hâbitat	Field recording	F1	F2	F3	F4
TINAMIFORMES Huxley, 1872								
TINAMIDAE Gray, 184O								
Crypturellus cinereus (Gmelin, 1789)	black inhambu	F	S	GRA			X	X
Crypturellus undulatus (Temminck, 1815)	jackfruit	F	S	GRA		X		X
Crypturellus strigulosus (Temminck, 1815)	inhambu-watch	F	S	GRA	X			
Crypturellus parvirostris (Wagler, 1827)	inhambu-chororó	F	S	GRA	X	X		
Crypturellus tataupa (Temminck, 1815)	inhambu-chinta	F	S	GRA			X	
GALLIFORMES Linnaeus, 1758								
CRACIDAE Rafinesque, 1815								
Penelope superciliaris Temminck, 1815	jacupemba	F	C	OBS	X			
PELECANIFORMES Sharpe, 1891								
ARDEIDAE Leach, i820								
Ardea alba Linnaeus, 1758	great egret	C	A	OBS	X			X
Pilherodius pileatus (Boddaert, 1783)	heron	C	S	FOT			X	
CATHARTIFORMES Seebohm, 189O								
CATHARTIDAE Lafresnaye, 1839								
Cathartes aura (Linnaeus, 1758)	red-headed vulture	D	P	FOT			X	
Coragyps atratus (Bechstein, 1793)	black-headed vulture	D	P	FOT			X	X
ACCIPITRIFORMES Bonaparte, 1831								
ACCIPITRIDAE Vigors, 1824								
Accipiter poliogaster (Temminck, 1824)	tauató-pintado	C	C	FOT		X		
Ictinia plumbea (Gmelin, 1788)	sovi	C	P	FOT			X	
Rupornis magnirostris (Gmelin, 1788)	sparrowhawk	C	B	FOT				X
Buteo nitidus (Latham, 1790)	rock hawk	C	C	FOT			X	X
FALCONIFORMES Bonaparte, issi								
FALCONIDAE Leach, 182O								
Daptrius ater Vieillot, 1816	santa hawk	C	C	OBS	X			
Ibycter americanus (Boddaert, 1783)	grit	C	C	VOC		X		X
Caracara plancus (Miller, 1777)	caracarà	C	B	FOT			X	
Herpetotheres cachinnans (Linnaeus, 1758)	acauâ	C	B	FOT	X		X	
Falco sparverius Linnaeus, 1758	kiriquiri	C	P	FOT			X	
Falco rufigularis Daudin, 1800	kayak	C	B	FOT				X
Falco femoralis Temminck, 1822	collared hawk	C	B	FOT	X			
GRUIFORMES Bonaparte, 1854								
RALLIDAE Rafinesque, 1815								
Porphyrio martinica (Linnaeus, 1766)	blue-water chicken	O	A	OBS			X	
CHARADRIIFORMES Huxley, 1867								
CHARADRIIi Huxley, 1867								
CHARADRIIDAE Leach, 1820								

Vanellus chilensis (Molina, 1782)	kite	O	A	OBS		X		
SCOLAPACI Steijneger, 1885								
SCOLAPACIDAE Rafinesque, 1815								
Tringa solitaria Wilson, 1813	solitary sandpiper	O	A	FOT	X			X
JACANIDAE Chenu & Des Murs, 1854								
Jacana jacana (Linnaeus, 1766)	jackfruit	O	A	FOT		X		
COLUMBIFORMES Latham, 1790								
COLUMBIDAE Leach, 1820								
Columbina passerina (Linnaeus, 1758)	grey turtledove	G	P	FOT			X	X
Columbina talpacoti (Temminck, 1811)	turtledove	G	P	OBS		X		X
Patagioenas speciosa (Gmelin, 1789)	trumpet dove	O	B	FOT				X
Patagioenas cayennensis (Bonnaterre, 1792)	wood pigeon	O	B	FOT				X
Leptotila rufaxilla (Richard & Bernard, 1792)	jellyfish	O	S	GRA		X		
PSITTACIFORMES Wagler, 1830								
PSITTACIDAE Rafinesque, 1815								
Ara macao (Linnaeus, 1758)	araracanga	F	C	OBS			X	X
Ara chloropterus Gray, 1859	great scarlet macaw	F	C	OBS			X	X
Aratinga weddellii (Deville, 1851)	dirty-headed parakeet	F	C	FOT		X	X	X
Brotogeris cyanoptera (Pelzeln, 1870)	blue-winged parakeet	F	C	OBS			X	X
Amazona ochrocephala (Gmelin, 1788)	field parrot	F	C	FOT		X	X	
CUCULIFORMES Wagler, 1830								
CUCULIDAE Leach, 1820								
Piaya cayana (Linnaeus, 1766)	cat's soul	I	P	FOT	X			X
CROTOPHAGINAE Swainson, 1837								
Crotophaga ani Linnaeus, 1758	black anu	I	P	OBS			X	X
STRIGIFORMES Wagler, 1830								
STRIGIDAE Leach, 1820								
Megascops choliba (Vieillot, 1817)	barn owl	C	B	GRA				X
Strix huhula Daudin, 1800	black owl	C	S	FOT			X	
Athene cunicularia (Molina, 1782)	burrowing owl	I	P	FOT	X	X	X	
CAPRIMULGIFORMES Ridgway, 1881								
CAPRIMULGIDAE Vigors, 1825								
Nyctiphrynus ocellatus (Tschudi, 1844)	ocellated bacurau	I	B	FOT				X
APODIFORMES Peters, 1940								
TROCHILIDAE Vigors, 1825								
PHAETHORNITHINAE Jardine, 1833								
Phaethornis philippii (Bourcier, 1847)	yellow whitetail	N	S	OBS			X	X
TROCHILINAE Vigors, 1825								
Thalurania furcata (Gmelin, 1788)	green scissors-flower	N	B	FOT	X	X		X
Heliomaster longirostris (Audebert & Vieillot, 1801)	grey-black beak	N	B	FOT			X	X
TROGONIFORMES A. O. U., 1886								
TROGONIDAE Lesson, 1828								
Trogon viridis Linnaeus, 1766	yellow-bellied great spotted owl	F	C	FOT	X		X	X
Trogon curucui Linnaeus, 1766	red-bellied tamarin	F	C	FOT			X	
CORACIIFORMES Forbes, 1844								
ALCEDINIDAE Rafinesque, 1815								
Chloroceryle amazona (Latham, 1790)	green kingfisher	P	A	FOT		X		
GALBULIFORMES Fürbringer, 1888								
GALBULIDAE Vigors, 1825								
Brachygalba lugubris (Swainson, 1838)	black sorrel	I	B	GRA			X	
Galbula ruficauda Cuvier, 1816	ariramba-de-cauda-ruiva	I	S	GRA			X	
Galbula dea (Linnaeus, 1758)	ariramba-do-paraiso	I	S	FOT	X		X	
BUCCONIDAE Horsfield, 1821								
Monasa nigrifrons (Spix, 1824)	black cry	I	B	FOT			X	X
Monasa morphoeus (Hahn & Küster, 1823)	white-faced weeping owl	I	S	FOT	X	X		X
Chelidoptera tenebrosa (Pallas, 1782)	little vulture	I	B	FOT		X		
PICIFORMES Meyer & Wolf, 1810								

RAMPHASTIDAE Vigors, 1825								
Ramphastos tucanus Linnaeus, 1758	great white fronted toucan	F	C	FOT		X		X
Pteroglossus inscriptus Swainson, 1822	striped-billed goshawk	F	C	FOT			X	
Pteroglossus castanotis Gould, 1834	brown goshawk	F	C	FOT	X	X	X	
PICIDAE Leach, 1820								
Picumnus aurifrons Pelzeln, 1870	golden dwarf woodpecker	I	B	FOT	X			
Melanerpes candidus (Otto, 1796)	white woodpecker	I	B	FOT		X		
Melanerpes cruentatus (Boddaert, 1783)	red-headed benedict	I	B	FOT	X	X		X
Veniliornis affinis (Swainson, 1821)	red-headed woodpecker	I	S	FOT		X		X
Piculusflavigula (Boddaert, 1783)	pileated woodpecker	I	S	FOT				X
Campephilus melanoleucos (Gmelin, 1788)	red-cockaded woodpecker	I	B	FOT			X	
PASSERIFORMES Linnaeus, 1758								
TYRANNI Wetmore & Miller, 1926								
THAMNOPHILIDA Patterson, 1987								
THAMNOPHILIDAE Swainson, 1824								
THAMNOPHILINAE Swainson, 1824								
Thamnophilus doliatus (Linnaeus, 1764)	brooding	I	S	VOC		X		
FURNARIIDA Sibley, Ahlquist & Monroe, 1988								
FURNARIOIDEA Gray, 1840								
DENDROCOLAPTIDAE Gray, 1840								
DENDROCOLAPTINAE Gray, 1840								
Glyphorynchus spirurus (Vieillot, 1819)	wedge-billed arapaçu	I	S	FOT			X	X
Xiphorhynchus obsoletus (Lichtenstein, 1820)	striped arapaçu	I	S	FOT			X	X
Dendrocolaptes certhia (Boddaert, 1783)	barred arapaçu	I	S	FOT			X	
TYRANNIDA Wetmore & Miller, 1926								
COTINGOIDEA Bonaparte, 1849								
COTINGIDAE Bonaparte, 1849								
COTINGINAE Bonaparte, 1849								
Lipaugus vociferans (Wied, 1820)	cricrió	F	C	VOC	X			X
Querula purpurata (Statius Muller, 1776)	anambé-una	F	C	FOT			X	
TYRANNOIDEA Vigors, 1825								
TYRANNIDAE Vigors, 1825								
TYRANNINAE Vigors, 1825								
Myiarchus tyrannulus (Statius Muller, 1776)	rusty-tailed mary janes	O	B	FOT			X	
Myiozetetes cayanensis (Linnaeus, 1766)	ferrugineous-winged warbler	O	B	FOT				X
Tyrannus melancholicus Vieillot, 1819	suiriri	O	P	FOT		X	X	X
PASSERI Linnaeus, 1758								
PASSERIDA Linnaeus, 1758								
HIRUNDINIDAE Rafinesque, 1815								
Progne tapera (Vieillot, 1817)	field swallow	I	P	OBS			X	X
TURDIDAE Rafinesque, 1815								
Turdus albicollis Vieillot, 1818	collared wisp	O	S	GRA			X	
THRAUPIDAE Cabanis, 1847								
Saltator maximus (Statius Muller, 1776)	viola seasoning	O	B	FOT		X		X
Ramphocelus carbo (Pallas, 1764)	red kite	O	B	OBS		X		X
Tangara velia (Linnaeus, 1758)	diamond skirt	O	B	FOT				X
Tangara episcopus (Linnaeus, 1766)	amazonian sandpiper	O	B	OBS				X
Tangarapalmarum (Wied, 1823)	Coconut sandpiper	O	B	OBS				X
Tangara nigrocincta (Bonaparte, 1838)	masked-skirt	O	B	FOT				X
Dacnis cayana (Linnaeus, 1766)	blue skirt	O	B	FOT				X
EMBERIZIDAE Vigors, 1825								
Ammodramus aurifrons (Spix, 1825)	field leafhopper	G	P	FOT				X
Sporophila angolensis (Linnaeus, 1766)	curio	G	B	OBS		X		
ICTERIDAE Vigors, 1825								
Psarocolius bifasciatus (Spix, 1824)	japuaçu	O	B	FOT			X	
Cacicus cela (Linnaeus, 1758)	shale	O	B	OBS		X	X	X
Sturnella militaris (Linnaeus, 1758)	police-ingles-of-the-north	G	P	FOT	X		X	

APPENDIX II

Trogon viridis (Yellow-bellied Surucuá-grande); Photo: Ângela Dias

Pteroglossus castanotis (Araçari-castanho); Photo: Ângela Dias

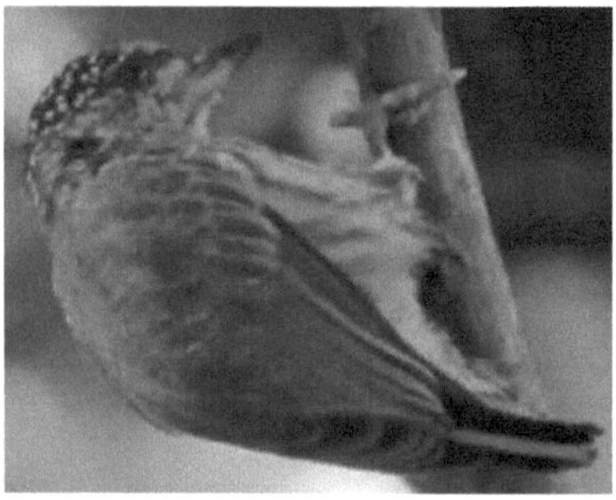

Picumnus aurifrons (Golden Dwarf Woodpecker); Photo: Ângela Dias

Melanerpes cruentatus (Red-fronted Benedict); Photo: Ângela Dias

Amazona ochrocephala (pampas parrot); Photo: Ângela Dias

Aratinga weddellii (Juicy-faced Parakeet); Photo: Ângela Dias

*Campephilus melanoleucos (*Red-cockaded Woodpecker); Photo: Ângela Dias

*Pilherodius pileatus (*Heron); Photo: Ângela Dias

*Trogon curucui (*Red-bellied Surucuá); Photo: Ângela Dias

Herpetotheres cachinnans (Acauã); Photo: Ângela Dias

*Crotophaga ani (*Black Anu); Photo: Ângela Dias

Sturnella militaris; Photo: Ângela Dias

Cathartes aura (Red-headed vulture); Photo: Ângela Dias

Athene cunicularia (Burrowing Owl); Photo: Ângela Dias

Columbina passerina (Grey turtledove); Photo: Ângela Dias

Tringa solitaria (Solitary Sandpiper); Photo: Ângela Dias

Melanerpes candidus (White Woodpecker); Photo: Ângela Dias

Chloroceyle amazona (Green kingfisher); Photo: Ângela Dias

MIX
Papier aus verantwortungsvollen Quellen
Paper from responsible sources
FSC® C105338

Printed by Books on Demand GmbH, Norderstedt / Germany